Mensa
The High IQ Society

MATHS TESTS

THIS IS A CARLTON BOOK

Published in 2016 by
Carlton Books
20 Mortimer Street
London W1T 3JW

ISBN: 978-1-78097-517-7

All images: © iStockphoto & Shutterstock

Printed in China

MATHS TESTS

CHALLENGE YOUR NUMERICAL POWERS
WITH MENSA-DERIVED CHALLENGES

CARLTON
BOOKS

What is Mensa?

Mensa is the international society for people with a high IQ.
We have more than 100,000 members in over 40 countries worldwide.

The society's aims are:
    to identify and foster human intelligence for the benefit of humanity
    to encourage research in the nature, characteristics, and uses of intelligence
    to provide a stimulating intellectual and social environment for its members

Anyone with an IQ score in the top two per cent of population is eligible to
become a member of Mensa – are you the 'one in 50' we've been looking for?

Mensa membership offers an excellent range of benefits:
    Networking and social activities nationally and around the world
    Special Interest Groups – hundreds of chances to pursue your hobbies
    and interests – from art to zoology!
    Monthly members' magazine and regional newsletters
    Local meetings – from games challenges to food and drink
    National and international weekend gatherings and conferences
    Intellectually stimulating lectures and seminars
    Access to the worldwide SIGHT network for travellers and hosts

For more information about Mensa: www.mensa.org, or

British Mensa Ltd.,
St John's House,
St John's Square,
Wolverhampton
WV2 4AH
Telephone: +44 (0) 1902 772771
E-mail: enquiries@mensa.org.uk
www.mensa.org.uk

4

CONTENTS

6

INTRODUCTION

INTRODUCTION

Puzzles are as old as humankind. It's inevitable – it's the way we think. Our brains make sense of the world around us by looking at the pieces that combine to make up our environment. Each piece is then compared to everything else we have encountered. We compare it by shape, size, colour, textures, a thousand different qualities, and place it into the mental categories it seems to belong to. Then also consider other nearby objects, and examine what we know about them, to give context. We keep on following this web of connections until we have enough understanding of the object of our attention to allow us to proceed in the current situation. We may never have seen a larch before, but we can still identify it as a tree. Most of the time, just basic recognition is good enough, but every time we perceive an object, it is cross-referenced, analysed, pinned down – puzzled out.

This capacity for logical analysis – for reason – is one of the greatest tools in our mental arsenal, on a par with creativity and lateral induction. Without it, science would be non-existent, and mathematics no more than a shorthand for counting items. In fact, although we might have made it out of the caves, we wouldn't have got far.

Furthermore, we automatically compare ourselves to each other – we place ourselves in mental boxes along with everything else. We like to know where we stand. It gives us an instinctive urge to compete, both against our previous bests and against each other. Experience, flexibility and strength are acquired through pushing personal boundaries, and that's as true of the mind as it is of the body. Deduction is something that we derive satisfaction and worth from, part of the complex blend of factors that goes into making up our self-image. We get a very pleasurable sense of achievement from succeeding at something, particularly if we suspected it might be too hard for us.

The brain gives meaning and structure to the world through analysis, pattern recognition, and logical deduction – and our urge to measure and test ourselves is an unavoidable reflex that results from that. So what could be more natural than spending time puzzling?

The dawn of puzzling

The urge to solve puzzles appears to be a universal human constant. They can be found in every culture, and in every time that we have good archaeological evidence for. The earliest material uncovered so far that is indisputably a puzzle has been dated to a little after 2000BC – and the first true writing we know of only dates back to 2600BC. The puzzle text is recorded on a writing tablet, preserved from ancient Babylonia. It is a mathematical puzzle based around working out the sides of a triangle.

Other puzzles from around the same time have also been discovered. The Rhind Papyrus from ancient Egypt describes a puzzle that is almost certainly a precursor of the traditional English riddle "As I Was Going to St. Ives." In the Rhind Papyrus, a puzzle is constructed around the clearly unreal situation of seven houses, each containing seven cats – and every cat kills seven mice that themselves had each consumed seven ears of millet.

In a similar foreshadowing, a set of very early puzzle jugs – Phoenician work from around 1700BC, found in Cyprus – echo designs that were to become popular in medieval Europe. These particular jugs, belonging to a broad category known as Askoi, had to be filled from the bottom. This form of trick vessel would later become known as a Cadogan Teapot. These devices have no lid, and have to be filled through a hole in the base. Because the hole funnels to a point inside the vessel, it can be filled to about half-way without spilling when it is turned back upright.

Earlier finds do exist, but so much context is lost down through the years that it can be difficult to be certain that the creators were thinking of puzzles specifically, or just of mathematical demonstrations. A set of ancient Babylonian tablets showing geometric progressions – mathematical sequences – is thought to be from 2300BC. One of the very first mathematical finds though, thought to possibly be from as far back as 2700 BC, is a set of stone balls carved into the shapes of the Platonic solids. These are regular convex polyhedrons – three-dimensional solid shapes made up solely of identical regular polygons. The most familiar is the basic cube, made up of six squares, but there are just four others – the tetrahedron, made up of four equilateral triangles; the octahedron, made up of eight equilateral triangles; the dodecahedron, made from twelve pentagons, and the icosahedron, made of twenty equilateral triangles.

There's no way now of knowing whether the carvings were teaching aids, puzzle or game tools, demonstrations of a theory, artistic constructions or even religious icons. The fact they exist at all however shows that someone had previously spent time working out a significant abstract mathematical puzzle – discovering which regular convex polyhedrons could exist.

The first great labyrinth

One of the greatest physical puzzles ever engineered comes from the same time period. The Egyptian Pharaoh Amenemhet III constructed a funerary pyramid with a huge temple complex around it in the form of an incredible labyrinth. Designed to guard the Pharaoh's mummy and treasures from disturbance or robbery, the labyrinth was so lavish and cunning that it is said to have been both the inspiration and template for the famous labyrinth that Daedalus built at Knossos for King Minos of Crete – the one that supposedly contained the Minotaur.

A history of puzzling

Coming forward in time, the evidence for the variety and complexity of puzzles gets ever stronger – an inevitable fact of archaeological and historical research. Greek legend claims that numbered dice were invented at the siege of Troy around 1200BC. We know that there was a craze for lateral thinking puzzles and logical dilemmas in the Greek culture from the 5th to 3rd centuries BC. A lot of very important mathematical work also took place in Greece from the middle of the first millennium BC, moving across to Rome during the first centuries AD. At the same time, the Chinese were playing with numerical puzzles and oddities, most famously the magic square, which they called Lo Shu (River Map), and also doing more strong mathematical work.

Puzzles and puzzle-like games that survive through to modern times get more common as we get closer to modern times, naturally. The game of Go arose in China some time around 500 BC, spreading to Japan a thousand years later – it is still an important sport there. At the same time, Chess was first appearing, either in India (Chaturanga), China (Xiang-qi), or both. Puzzle rings that you have to find out how to separate also appeared in China, possibly in the 3rd century AD, as did Snakes & Ladders, around 700AD.

The first known reference to a game played with cards is in 969AD, in records reporting the activities of the Chinese Emperor Mu-tsung. These are not thought to be the playing cards now familiar in the west, however – it seems likely that those arose in Persia during the 11th or 12th century AD. The physical puzzle Solitaire is first reported in 1697AD. As the eighteenth century gave way to the nineteenth, the forces of the industrial revolution really started to transform the way that ideas propagated, and the puzzle world exploded. Some of the more notable highlights include the invention of the jigsaw puzzle by John Spilsbury in 1767; Tic-Tac-Toe's first formal discussion in 1820, by Charles Babbage; poker first appearing around 1830 in the USA; Lucas inventing the Tower of Hanoi puzzle in 1883; the first crossword appearing in New York World on December 21, 1913, created by Arthur Wynne; Erno Rubik's invention of his Cube in 1974; and the invention of Sudoku in 1979 for Dell Magazines by Howard Garns, an American, who first called it "Number Place".

Good for the brain?

It turns out that it's a good thing puzzles are such an important part of the human psyche. Recent advances in the scientific fields of neurology and cognitive psychology have hammered home the significance of puzzles and mental exercise like never before.

We now understand that the brain continually builds, shapes and organises itself all through our lives. It is the only organ to be able to do so. Previously, we had assumed that the brain was constructed to optimise infant development, but the truth is that it continually rewrites its own operating instructions. It can route around physical damage, maximise its efficiency in dealing with commonly encountered situations and procedures, and alter its very

structure in response to our experiences. This incredible flexibility is referred to as plasticity.

The most important implication of plasticity is that our mental abilities and cognitive fitness can be exercised at any age. Just like the muscles of the body, our minds can respond to exercise, allowing us to be more retentive and mentally fitter. Our early lives are the most important time, of course. Infants develop almost twice as many synapses – the mental connections that are the building-blocks of the mind – as we retain as adults, to make sure that every experience can be learnt from and given its own space in the developing mental structure. The first thirty-six months are particularly vital, the ones which will shape the patterns of our intellect, character and socialisation for life. A good education through to adulthood – stretching the brain right through childhood – is one of the strongest indicators of late-life mental health, particularly when followed with a mentally challenging working life.

Just as importantly however, there is little difference between the brain at the age of 25 and the age of 75. As time passes, the brain optimises itself for the lifestyle we feed it. Circuits that are hardly ever used get re-adapted to offer greater efficiency in tasks we regularly use. Just as our body maximises available energy by removing muscle we don't use, the brain removes mental tone we're never stretching – and in the same way that working out can build up muscle, so mental exercise can restore a "fit" mind.

Puzzle solving and brain growth

A surprising amount of mental decline in elders is now thought to be down to insufficient mental exercise. Where severe mental decline occurs, it is usually linked to the tissue damage of Alzheimer's Disease – although there is now even evidence that strong mental exercise lets the brain route around even Alzheimer's damage, lessening impairment. In other cases, where there is no organic damage, the main cause is disuse. Despite old assumptions, we do not significantly lose huge swathes of brain cells as we age. Better still, mental strength that has been allowed to atrophy may be rebuilt.

Research projects across the world have discovered strong patterns linking highly lucid venerable people. These include above-average education, acceptance of change, satisfying personal accomplishments, physical exercise, a clever spouse, and a strong engagement with life, including reading, social activity, travel, keeping up with new ideas, and regularly solving puzzles.

Not all the things we assume to be engagement are actually helpful, however. Useful intellectual pursuits are the actively stimulating ones – such as solving jigsaws, crosswords and other puzzles, playing chess, and reading books that stimulate the imagination or require some mental effort to properly digest. However, passive intellectual pursuits may actually hasten the mind's decay. Watching television is the most damaging such pastime, but surprisingly anything that makes you "switch off" mentally can also be harmful, such as listening to certain types of music, reading very low-content magazines and even getting most of your social exposure on the telephone. For social interaction to be helpful, it may really need to be face to face.

The Columbia Study

A team of researchers from Columbia University in New York tracked more than 1,750 pensioners from the northern Manhattan region over a period of seven years. The subjects underwent periodic medical and psychological examination to assess both their mental health and the physical condition of their brains. Participants also provided the researchers with detailed information regarding their daily activities. The study found that even when you remove education and career attainment from the equation, leisure activity significantly reduced the risk of dementia.

The study's author, Dr Yaakov Stern, found that "Even when controlling for factors like ethnic group, education and occupation, subjects with high leisure activity had 38% less risk of developing dementia." Activities were broken into three categories: physical, social and intellectual. Each one was found to be beneficial, but the greatest protection came from intellectual pursuits. The more activity, the greater the protection – the cumulative benefit of each separate leisure pursuit was found to be 8%. Stern also found that leisure activity helped to prevent the physical damage caused by Alzheimer's from actually manifesting as dementia:

> "Our study suggests that aspects of life experience supply a set of skills or repertoires that allow an individual to cope with progressing Alzheimer's Disease pathology for a longer time before the disease becomes clinically apparent. Maintaining intellectual and social engagement through participation in everyday activities seems to buffer healthy individuals against cognitive decline in later life."

Staying lucid

There is strong evidence to back Stern's conclusion. Dr David Bennett of the Rush Alzheimer's Disease Centre in Chicago led a study that evaluated a group of venerable participants on a yearly basis, and then after death examined their donated brains for signs of Alzheimer's. The participants all led active lives mentally, socially and physically, and none of them suffered from dementia at the time of their death. It was discovered that more than a third of the participants had sufficient brain-tissue damage to warrant diagnosis of Alzheimer's Disease, including serious lesions in the brain tissue. This group had recorded lower scores than other participants in episodic memory tests – remembering story episodes, for example – but performed identically in cognitive function and reasoning tests.

A similar study took place with the aid of the nuns of the Order of the School Sisters of Notre Dame. The Order boasts a long average lifespan – 85 years – and came to the attention of researchers when it became clear that its members did not seem to suffer from any dementia either. The distinguishing key about the Order is that the nuns shun idleness and mental vacuity, taking particular effort to remain mentally active. All sorts of pursuits are encouraged, such as solving puzzles, playing challenging games, writing, holding seminars on current affairs, knitting and engaging with local government. As before, there was plenty of evidence of the physical damage associated with Alzheimer's Disease, but none of the mental damage that usually accompanied it, even in some nonagenarian participants.

Mental repair

Other studies have also tried to enumerate the benefits of mental activity. A massive group study led by Michael Valenzuela from the University of New South Wales' School of Psychiatry tracked data from almost 30,000 people worldwide. The results were clear – as well as indicating the same clear relationship previously found between schooling, career and mental health, people of all backgrounds whose daily lives include a high degree of mental stimulation are 46% less likely to suffer dementia. This holds true even for people who take up mentally challenging activities as they get older – if you use your mind, the brain still adapts to protect it. If you do not use it, the brain lets it falter.

Puzzle solving techniques

Puzzle solving is more of an art than a science. It requires mental flexibility, a little understanding of the underlying principles and possibilities, and sometimes a little intuition. It is often said of crosswords that you have to learn the writer's style to get really good at his or her puzzles, but the same thing applies to most other puzzle types to a certain extent, and that includes the many and various kinds you'll find in this book.

Sequence puzzles

Sequence puzzles challenge you to find a missing value or item, or to complete a pattern according to the correct underlying design. In this type of puzzle, you are provided with enough previous entries in the sequence that the underlying logic can be worked out. Once the sequence is understood, the missing entry can be calculated. When the patterns are simple, the sequence will be readily visible to the naked eye. It is not hard to figure out that the next term in the sequence 1, 2, 4, 8, 16, ? is going to be a further doubling to 32. Numerical sequences are just the expression of a mathematical formula however, and can therefore get almost infinitely complex.

Proper recreational puzzles stay firmly within the bounds of human ability, of course. With the more complex puzzles, the best approach is often to calculate the differences between successive terms in the sequences, and look for patterns in the way that those differences are changing. You should also be aware that in some puzzles, the terms of a sequence may not necessarily represent single items. Different parts or digits of each term may progress according to different calculations. For example, the sequence 921, 642, 383, 164 is actually three simple sequences stuck together - 9, 6, 2, 0 ; 2, 4, 8, 16; and 1, 2, 3, 4. The next term will be -3325. Alternatively, in puzzles where the sequence terms are given as times, they may actually just represent the times they depict, but they might also be literal numbers, or pairs of numbers to be treated as totally different sequences, or even require conversion from hours:minutes to just minutes before the sequence becomes apparent.

For example, 11:14 in a puzzle might represent the time 11:14, or perhaps the time 23:14 – or the numbers 11 and 14, the numbers 23 and 14, the number 1114, the number 2314, or even the number 674 (11 * 60 minutes, with the remaining 14 minutes also added). As you can see, solving sequence puzzles requires a certain amount of trial and error as you test difference possibilities, as well as a degree of lateral thinking. It would be a very harsh puzzle setter who expected you to guess some

sort of sequence out of context however. So in the absence of a clue otherwise, 11:14 would be highly unlikely to represent 11 months and 14 days, or the value 11 in base 14, or even 11 hours and 14 minutes converted to seconds – unless it was given as 11:14:00, of course.

Letter-based sequences are all representational of course, as unlike numbers, letters have no underlying structure save as symbols. Once you deduce what the letters represent, the answer can be obvious. The sequence D, N, O, ? may seem abstract, until you think of months of the year in reverse order. In visual sequences – such as pattern grids – the sequence will always be there for you to see, and your task is to look for repeating patterns. As with number sequences, easy grids can be immediately apparent. In harder puzzles, the sequences can become significantly long, and often be presented in ways that make them difficult to identify. Puzzle setters love to start grids of this type from the bottom right-hand square, and then progress in spirals or in a back-and-forth pattern – sometimes even diagonally.

Odd-one-out problems are a specialised case of sequence pattern where you are given the elements of a sequence or related set, along with one item that breaks the sequence. Like other sequence puzzles, these can range from very easy to the near-impossible. Spotting the odd one in 2, 4, 6, 7, 8 is trivial. It would be almost impossible to guess the odd item from the set of B, F, H, N, O unless you already knew that the set in question was the physical elements on the second row of the standard periodic table. Even then, you might need a copy of the periodic table itself to notice that hydrogen, H, is on the first row. As with any other sequence problem, any odd-one-out should contain enough information in the puzzle, accompanying text and title to set the context for finding the correct answer. In the above case, a puzzle title along the lines of "An Elementary Puzzle" would probably be sufficient to make it fair game.

Equation puzzles

Equation puzzles are similar to sequences, but require a slightly different methodology. In these problems, you are given a set of mathematical calculations that contain one or more unknown terms. These may be represented as equations, as in the traditional form of $2x + 3y = 9$, or they may be presented visually, for example as two anvils and three iron bars on one side of a scale and nine horseshoes balancing on the other side of the scale.

For each Unknown – x, y, anvils, etc – you need one equation or other set of values before you can calculate a definitive answer. If these are lacking, you cannot get the problem down to just one possible solution. Take the equation above, $2x + 3y = 9$. There are two unknowns, and therefore many answers. For example, x can be 3 and y can be 1 – for x, $2 * 3 = 6$; for y, $3 * 1 = 3$, and overall, $6 + 3 = 9$ – but x can also be 1.5 and y can be 2... and an infinite range of other possibilities. So when solving equation puzzles, you need to consider all the equations together before you can solve the problem.

To return to our example equation above, if you also knew that $x + 2y = 7$, you could then begin to solve the puzzle. The key with equation problems is to get your equation down to containing just one unknown term, which then lets you get a value for that term, and in turn lets you get the value of the other unknown/s. So, for example, in our previous equations ($2x + 3y = 9$ and $x + 2y = 7$) you could manipulate one equation to work out what x actually represents in terms of y ("How many Y is each X?") in one equation, and then replace the x in the other equation with it's value in y, to get a calculation that just has y as the sole unknown factor. It's not as confusing as it sounds so long as you take it step by step:

We know that

x + 2y = 7

Any change made to both sides of an equation balances out, and so doesn't change the truth of the equation. For example, consider 2 + 2 = 4. If you add 1 to each side, the equation is still true. That is, 2 + 2 + 1 = 4 + 1. We can use this cancelling out to get x and y on opposite sides of the equation, which will let us represent x in terms of y:

x + 2y - 2y = 7 - 2y.

Now the + 2y - 2y cancels out:

x = 7 - 2y.

Now we know x is a way of saying "7-2y", we can replace it in the other equation. 2x + 3y = 9 becomes:

2 * (7 - 2y) + 3y = 9.

Note 2x means that x is in the equation twice, so our way of re-writing x as y needs to be doubled to stay accurate. Expanding that out:

(2 * 7) - (2 w* 2y) + 3y = 9, or
14 - 4y + 3y = 9.

The next step is to get just amounts of y on one side, and numbers on the other.

14 - 4y + 3y - 14 = 9 - 14.

In other words,

-4y + 3y = -5.

Now, -4 + 3 is -1, so:

-y = -5, and that means y=5.

Now you can go back to the first equation, x + 2y = 7, and replace y to find x.

x + (2 * 5) = 7
x + 10 = 7
x + 10 - 10 = 7 - 10
x = 7 - 10

and, finally.

x = -3.

As a last step, test your equations by replacing your number values for x and y in both at the same time, and making sure they balance correctly.

2x + 3y = 9 and x + 2y = 7.
(2 * -3) + (3 * 5) = 9 and -3 + (2 * 5) = 7
(-6 + 15) = 9; and (-3 + 10) = 7.
9 = 9 and 7 = 7.

The answers are correct.

Any equation-based puzzle you're presented with will contain enough information for you to work out the solution. If more than two terms are unknown, the technique is to use one equation to find one unknown as a value of the others, and then replace it in all the other equations. That gives you a new set of equations containing one less unknown term. You then repeat the process of working out an unknown again, until you finally get down to one unknown term and its numerical value. Then you replace the term you now know with its value in the equations for the level above to get the next term, and continue back on up like that. It's like a mathematical version of the old wooden Towers of Hanoi puzzle. As a final tip, remember that you should have one equation per unknown term, and that if one of your unknown variables is missing from an equation, the equation can be said to have 0 of that variable on either or both sides. That is, $4y + 2z = 8$ is the same as $0x + 4y + 2z = 8$.

Happy puzzling!

THE
PUZZLES

01 **461343** is the code for Tahiti. Using the same code can you work out the names of the following islands?

8653

38326

Answer see page 118

16343

20

Answer see page 118

02 Car A and car B set off from the same point, at the same time, to travel the same journey. Car A travels at 45 mph and car B travels at 35 mph. If car A stops after 70 miles, how long will it take car B to catch up?

A

B

What number should replace the question mark in the grid?

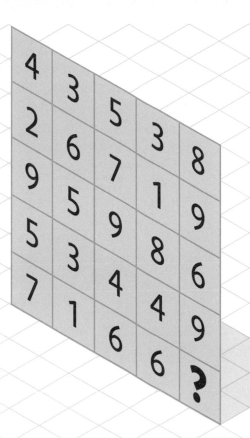

Answer see page 118

Answer see page 118

Assume you are using a basic calculator and apply the mathematical operations strictly in the order chosen. Replace each question mark with a mathematical sign, plus, minus, multiply and divide can each be used once only. In which order should they be used to score 13?

$3 \; ? \; 8 \; ? \; 7 \; ? \; 9 \; ? \; 2 = 13$

05 Clock A was correct at midnight. From that moment it began to lose three and a half minutes per hour. The clock stopped one and a half hours ago showing clock B. What is the correct time now? The clock runs for less than 24 hours.

A

B

Answer see page 118

Answer see page 118

06 Assume you are using a basic calculator and apply the mathematical operations strictly in the order chosen. Replace each question mark with a mathematical sign. Plus, minus, multiply and divide can each be used once only. What are the highest and lowest numbers you can possibly score?

9 ? 3 ? 2 ? 7 ? 4 = ◯

07

What number should replace the question mark?

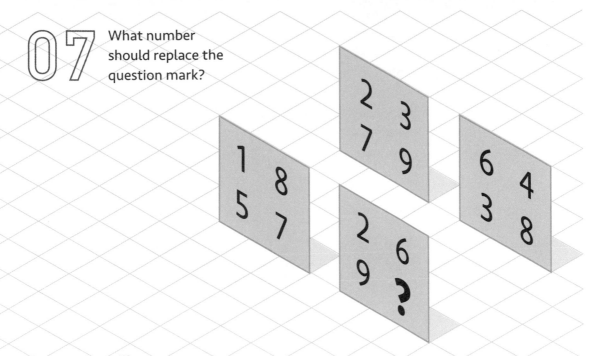

Answer see page 119

08

If J Y = 35, CG = 10 and LT = 32, what does BW = ?

Answer see page 119

09 A car and a motorcycle set off from the same point to travel the same journey. The car sets off two minutes before the motorcycle. If the car travels at 60 km/h and the motorcycle travels at 80 km/h, how many kilometres from the starting point will they draw level?

Answer see page 119

10 Clock A was correct at midnight. From that moment it began to lose four minutes per hour. The clock stopped three hours ago showing clock B. What is the correct time now? The clock runs for less than 24 hours.

A

B

Answer see page 119

Answer see page 119

11 What number should replace the question mark in the grid?

3	6	5	9
1	7	3	8
5	3	4	6
2	1	7	9
4	3	6	5
		9	
		2	**?**

12 The alphabet is written here but some letters are missing. Arrange the missing letters to give a word. What is it?

B C F G J K L M O

P Q R T U W X Y Z

Answer see page 119

Answer see page 119

13 Two vehicles set off from the same point to travel the same journey. The first vehicle sets off ten minutes before the second vehicle. If the first vehicle travels at 55 km/h and the second vehicle travels at 60 km/h, how many kilometres from the starting point will the two vehicles draw level?

26

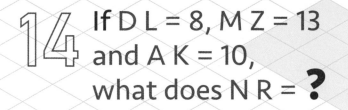

14 If D L = 8, M Z = 13 and A K = 10, what does N R = ?

Answer see page 119

Answer see page 119

15 What number should replace the question mark?

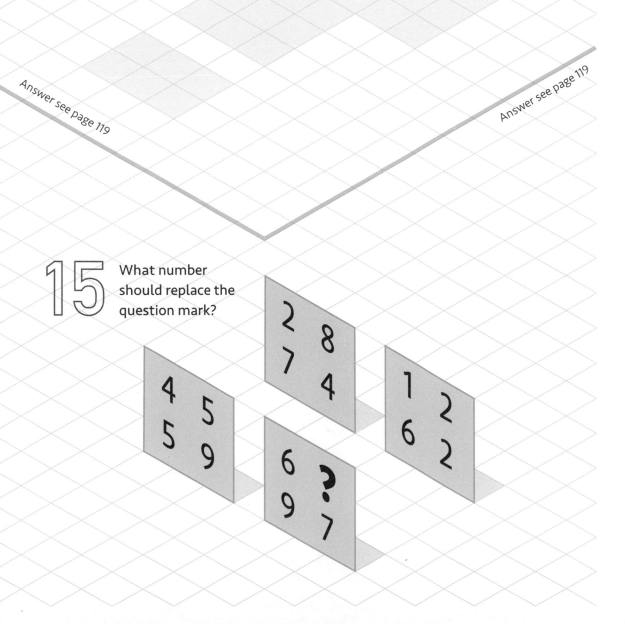

16

If A+M = 7, E−W = 0
and N−V=1,
what does H+Z = **?**

Answer see page 120

28

17

Assume you are using a basic
calculator and apply the
mathematical operations strictly
in the order chosen. Replace
each question mark with a
mathematical sign. Plus, minus,
multiply and divide can each
be used once only. How many
different ways are there to score 5?

Answer see page 120

(2) (?) (3) (?) (9) (?) (5) (?) (8) = (5)

18 An aeroplane covers its outward journey at 555 mph. It returns, over exactly the same distance at 370 mph. What is the aeroplane's average speed over the entire journey?

Answer see page 120

19 If CSF = 16, TAQ = 4, ZOL = 29 and HWM = 18, what does NER = **?**

Answer see page 120

Answer see page 120

20 Two cars set off from the same point, at the same time, to travel the same journey. The first car travels at 50 mph and the second car travels at 40 mph. If the first car stops after 90 miles, how many minutes will it take the second car to catch up?

21 A bus has travelled 60 miles at 50 mph. It started its journey with 8 gallons of fuel but its tank has been leaking throughout the journey and is now dry. The bus completes 25 miles per gallon. How many gallons of fuel does it leak per hour?

Answer see page 120

Answer see page 121

22 Add together three numbers each time to score 22. Each number can be used as many times as you wish. How many different combinations are there?

2   4   6   8   10   12   14

Answer see page 121

500 gallons

A fire engine travels 9 miles to a fire at a speed of 40mph. Its tank holds 500 gallons of water but has been leaking throughout the journey at a rate of 20 gallons per hour. If the fire engine requires 496 gallons of water to extinguish the fire, will it succeed?

What number should appear next in this sequence?

Answer see page 121

Answer see page 121

The sum of each two adjacent squares gives the number above. What number should replace the question mark?

26 A spaceship covers its journey to Earth at 735 mph. It returns, over exactly the same distance at 980 mph. What is the spaceship's average speed over the entire journey?

Answer see page 121

34

27 A factory recycles sheets of paper for use in its offices. Six used sheets of paper are needed to make each new sheet. If there are 2331 used sheets of paper, how many new sheets can possibly be made in total?

Answer see page 122

28 A 220 yard long train, travelling at 30 mph, enters a 3 mile long tunnel. How long will elapse between the moment the front of the train enters the tunnel and the moment the end of the train clears the tunnel?

Answer see page 122

Throw three darts at this board to score 70. How many different combinations are there? Every dart scores.

Answer see page 122

36

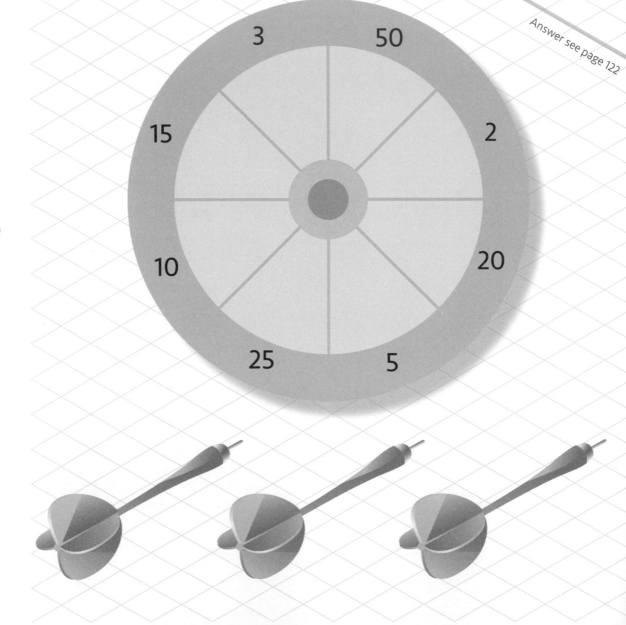

3   50

15   2

10   20

25   5

30 A collection raises £18.08. It is made up of four different denominations of coins and the largest denomination is £1. There is exactly the same number of each coin. How many of each coin is there and what are their values? The list of possible coins in the currency is:

1p, 2p, 5p, 10p, 20p, 50p £1.

Answer see page 122

31 A car has travelled 40 miles at 30 mph. It started its journey with 10 gallons of fuel but its tank has been leaking throughout the journey and is now dry. The car completes 30 miles per gallon. How many gallons of fuel does it leak per hour?

Answer see page 122

32 A 550 yard long train, travelling at 90 mph, enters a 2 mile long tunnel. How many seconds will elapse between the moment the front of the train enters the tunnel and the moment the end of the train clears the tunnel?

Answer see page 122

33 What number should appear next in this sequence?

Answer see page 123

(10) (30) (70) (130) (210) (**?**)

34 A ship is battling against the tide to safety. The ship uses 8 gallons of fuel every hour and sails at 22 mph in still conditions. The ship is 39 miles from safety and the flow against it is 7 mph. If the ship has 21 gallons of fuel remaining, will it reach safety?

Answer see page 123

A fire engine travels 7 miles to a fire at a speed of 42mph. Its tank holds 500 gallons of water but has been leaking throughout the journey at a rate of 22.5 gallons per hour. If the fire engine uses 495 gallons of water extinguishing the fire, how much water will it have left over?

Answer see page 123

Answer see page 123

What number is missing in this sequence?

5   6   11   17   45   ?   118

37 Complete the square with the letters of BEACH so that no row, column or diagonal line of any length contains the same letter more than once. What letter must replace the question mark?

B

E

A

H ? C

Answer see page 123

38

A group on a five day hiking holiday cover two fifths of the total distance on the first day. The next day they cover one quarter of what is left. The following day they cover two fifths of the remainder and on the fourth day half of the remaining distance. The group now have 15 miles left. How many miles have they walked?

Answer see page 123

Answer see page 123

39

A cyclist undertakes a 144-mile cycle ride for charity. On the first day he covers one third of the total distance. The next day he covers one third of what is left. The following day he covers one quarter of the remainder and on the fourth day half of the remaining distance. How many miles will he need to cycle on day 5 to get to the end of the ride?

What number should appear next in this sequence?

(1944) (648) (108) (12) (**?**)

Answer see page 123

Answer see page 124

The sum of each two adjacent squares gives the number above. What number should replace the question mark?

48

?

79   11

55

54

36

43

42

A factory recycles paper plates for use in its canteen. Nine used plates are needed to make each new plate. If there are 1481 used plates, how many new plates can possibly be made in total?

Answer see page 124

44

43

What numbers should replace the question marks?

Answer see page 124

10	200	30
11	165	26
12	120	22
13	?	?

In which direction
should the missing
arrow point?

Answer see page 124

In which direction
should the missing
arrow point?

Answer see page 124

46

46 What number should appear next in this sequence?

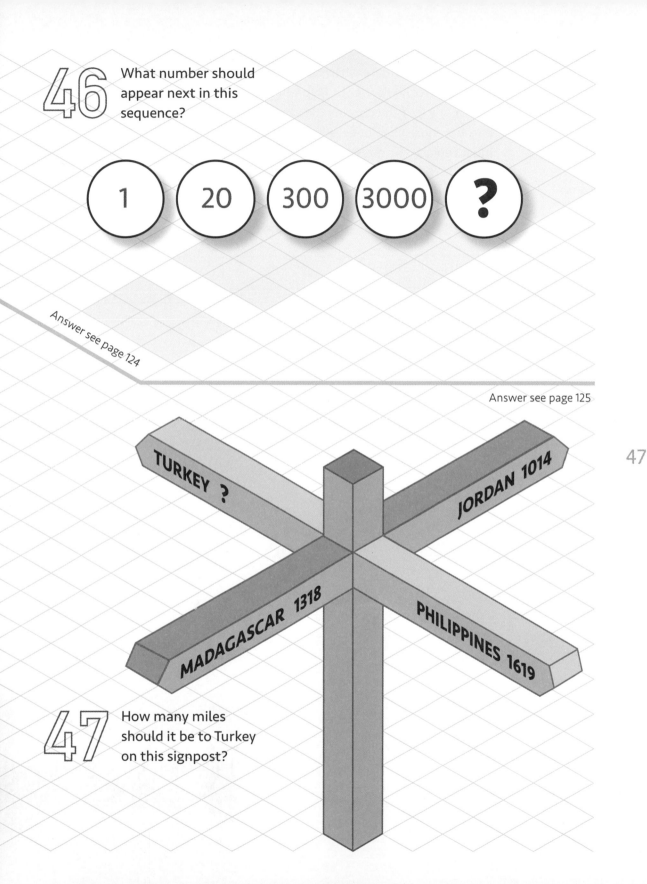

(1) (20) (300) (3000) (?)

Answer see page 124

Answer see page 125

TURKEY ?

JORDAN 1014

MADAGASCAR 1318

PHILIPPINES 1619

47 How many miles should it be to Turkey on this signpost?

48 Complete the square with the numbers 1 to 5 so that no row, column or diagonal line of any length contains the same number more than once. What number must replace the question mark?

Answer see page 125

1				
			2	
		3		
5				
			4	
				?

49 What number should appear next in this sequence?

Answer see page 125

10   12   16   22   30   ?

What should be the value of the fourth column?

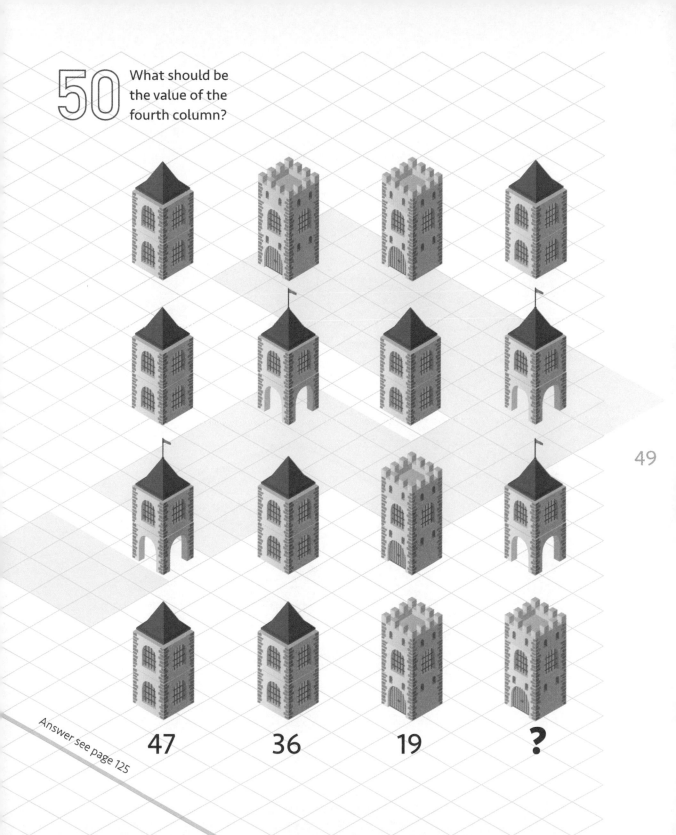

47   36   19   ?

51 Car A and car B set off from the same point, at the same time, to travel the same 115 mile journey. If car A travels at 48 mph and car B travels at 40 mph, what will be the difference in their arrival times?

50

A

B

Answer see page 125

52 What number should replace the question mark?

?	E	A	4
K			V
L			M
8	J	R	9

Answer see page 126

53 What letter should replace the question mark?

Answer see page 126

DLH   PUE   BTR   KYN   MWJ   E?A

Answer see page 126

52

54 There are seven items on a deli counter. The pastries are between the pickles and the curries. The salads are next to the samosas. There are two items between the ham and the curries and the ham is between the pickles and the cheese. The pastries are in the exact centre and the salads are at the far end of the counter. What is the order of the seven items?

55 The sum of each two adjacent squares gives the number above. What number should replace the question mark?

19

?

34 49

24

18

3

Answer see page 126

56 Which number is the odd one out?

313 428

236 339

259

224 188

Answer see page 126

57 What number is missing in this sequence?

(2) (12) (30) (56) (?) (132)

Answer see page 126

58 A coach has been travelling for four hours. In the first hour it covered one third of the total distance. The next hour it covered one third of what was left. The following hour it covered one quarter of the remainder and in the fourth hour half of the remaining distance. The coach now has 25 miles left to its destination. How many miles has it travelled?

Answer see page 126

In which direction
should the missing
arrow point?

Answer see page 127

?

A factory recycles cups for use in its canteen. Eight used cups are needed to make each new cup. If there are 736 used cups, how many new cups can possibly be made in total?

Answer see page 127

Answer see page 127

Assume you are using a basic calculator and apply the mathematical operations strictly in the order chosen. Replace each question mark with a mathematical sign. Plus, minus, multiply and divide can each be used once only. In which order should they be used to score 3?

$$4 \; ? \; 5 \; ? \; 9 \; ? \; 8 \; ? \; 7 = 3$$

What number
should replace the
question mark?

7

2

?

4

5

9

2

1

3

Answer see page 127

57

A man walks south for 4 miles
Then east for 3 miles
Then north for 1 mile
Then west for 1 mile
Then north for 3 miles

In which direction and for how far
should he walk to return to his
starting point?

Answer see page 127

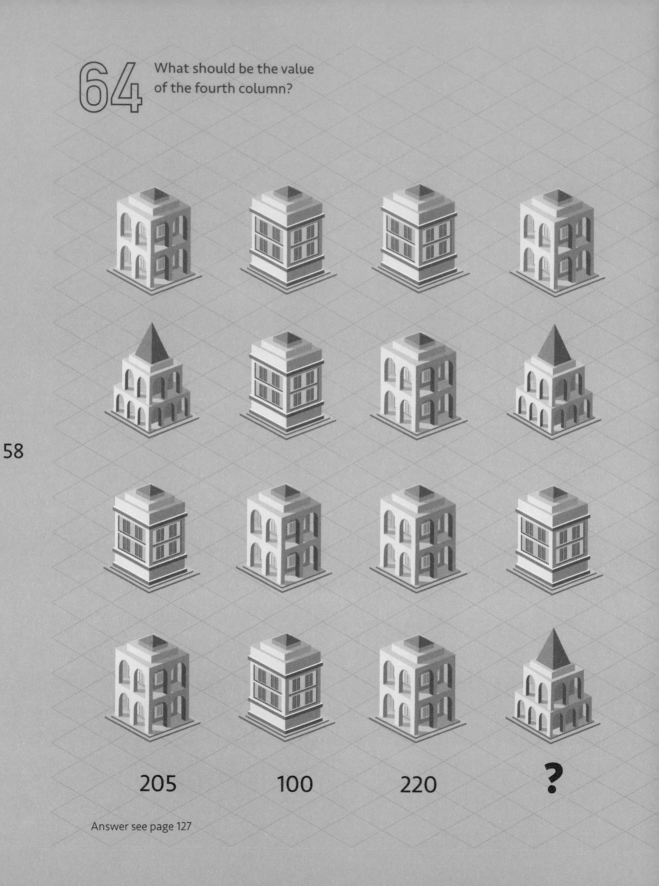

64 What should be the value of the fourth column?

205   100   220   ?

Answer see page 127

65 A ship is battling against a strong tide to safety. It uses eight gallons of fuel every hour and sails at 16 mph in still conditions. The ship is 84 miles from safety and the flow against it is 7 mph. The ship has 75 gallons of fuel left. How much will it have to spare when it reaches the shore?

Answer see page 127

Answer see page 127

59

66 What number should replace both question marks?

| 1 | 2 |
| 8 | 4 |

| 1 | 5 |
| 6 | 9 |

| 0 | 9 |
| 2 | 7 |

| ? | ? |
| 3 | 8 |

Clock A was correct at midnight.
From that moment it began to lose
three and a half minutes per hour.
The clock stopped an hour ago
showing clock B. What is the correct
time now? The clock runs for less
than 24 hours.

B

A

Answer see page 128

Answer see page 128

What number
should replace the
question mark?

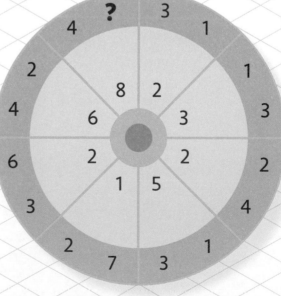

What numbers should replace the question mark?

12	4	9
57	19	54
33	11	30
48	16	45
27	?	?

Answer see page 128

70

A cyclist rides from one town to another. On the first day she covers one quarter of the total distance. The next day she covers one third of what is left. The following day she covers one quarter of the remainder and on the fourth day half of the remaining distance. The cyclist now has 25 miles left. How many miles has she travelled?

Answer see page 128

71 A 440 yard long tram, travelling at 40 mph, enters a tunnel of one and a half a miles in length. How long will elapse between the moment the front of the train enters the tunnel and the moment the end of the train clears the tunnel?

Answer see page 128

72 From a group of office workers, three times as many people choose attending a gym as choose swimming to keep fit. Six more people choose walking than choose swimming and three less people choose jogging than walking. 7 people choose jogging as their favourite activity. How many people choose each of the other activities?

Answer see page 128

73 What should be the value of the fourth row?

110

63

Answer see page 129

350

190

?

74 A cash till contains £15.48. It is made up of four different denominations of coins and the largest denomination is £1. There is exactly the same number of each coin. How many of each coin is there and what are their values?

Answer see page 129

64

75 A fire engine travels five miles to a fire at 30 mph. Its tank holds 500 gallons of water but has been leaking throughout the journey at a rate of 22.5 gallons per hour. How many gallons of water will the fire engine have available to put out the fire?

Answer see page 129

500 gallons

76

Here is an unusual safe. Each of the buttons must be pressed once only in the correct order to reach the centre X and open the safe. The number of moves and direction to move is marked on each button. Which button is the first you must press?

3 E	3 E	4 S	1 W	2 S
2 S	1 S	2 S	1 W	3 W
1 N	1 W	**X**	2 S	1 N
4 E	1 S	1 W	1 N	1 S
4 N	4 N	2 N	3 N	4 W

Answer see page 129

77 Throw three darts at this board to score 25. How many different combinations are there? Every dart lands in the board and no dart falls out.

Answer see page 129

What should be the value
of the fourth column?

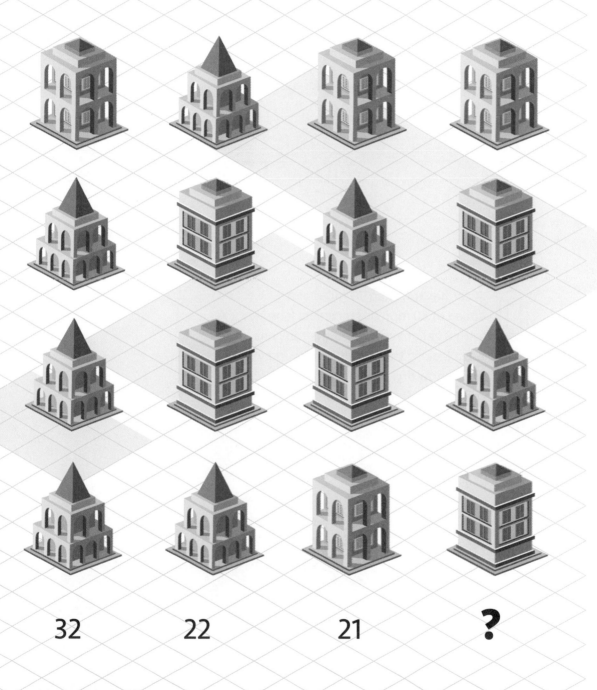

32   22   21   ?

Answer see page 129

What number should replace the question mark?

8 28

12 42

2 7

18 63

10 $?$

Answer see page 129

Answer see page 130

80 A car has travelled 30 miles at a speed of 50 mph. It started its journey with 8 gallons of fuel but its tank has been leaking throughout the journey and is now dry. The car completes 25 miles per gallon. How much fuel does it leak per hour?

81 In an amateur soccer team's first 15 games, the average number of goals per game was three. After a further 30 games the average goals per game increased to five. What was the average number of goals per game in the last 30 games only?

Answer see page 130

Answer see page 130

82 Add together three numbers each time to score 14. How many different combinations are there? Each number can be used as many times as you wish.

2 4 6 8 10

83 A coach and a car set off from the same point to travel the same journey. The coach leaves nine minutes before the car. If the coach travels at 50 km/h and the car travels at 80 km/h, how many kilometres from the starting point will they draw level?

Answer see page 130

Answer see page 130

84 A certain month has five **Sundays** and the first **Saturday** of the month is the 7th.

On what day does the 30th fall?

What will be the date of the third **Monday** of the month?

How many **Fridays** are there in the month?

On what day does the 11th fall?

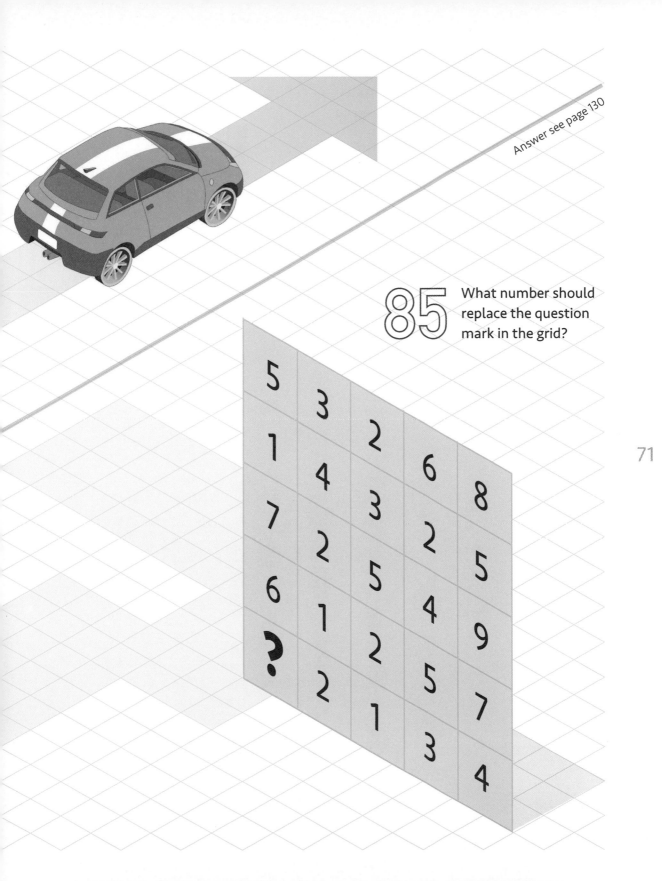

Answer see page 130

85 What number should replace the question mark in the grid?

5	3	2	6	8
1	4	3	2	5
7	2	5	4	9
6	1	2	5	7
?	2	1	3	4

Using only the numbers and signs given, create a sum where both sides are equal.

(6) (13) (17) (25) (+) (÷) (√) (=) ( )

Answer see page 130

Which symbol should replace the question mark to continue the sequence?

Answer see page 130

72

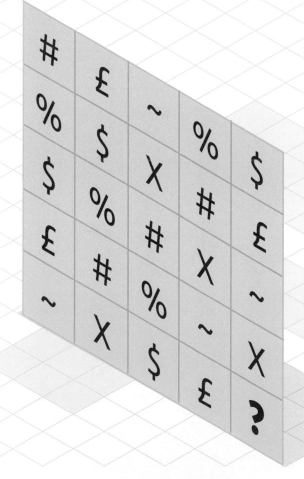

88 A helicopter covers its outward journey at 250 mph. It returns, over exactly the same distance at 166.6666 mph. What is the helicopter's average speed over the entire journey?

Answer see page 131

Answer see page 131

89 284561 is the code for CHAPEL, 67539 is the code for PEACH and 3867 is the code for LEAP.

Which word is 874 in code?

The number of miles to deserts is shown on this signpost. How many miles should it be to the Gobi?

KALAHARI 20

ARABIAN 15

MOJAVE 18

GOBI ?

Answer see page 131

How many dashes should
be contained on each row
of the final box?

92

A hiker walks from one town to another. On the first day he covers two fifths of the total distance. The next day he covers one third of what is left. The following day he covers one quarter of the remainder and on the fourth day half of the remaining distance. He now has 12 miles left. How far has he walked?

Answer see page 131

93

The following numbers watch sports on television. How many watch Hockey?

Answer see page 132

Boxing   – 11

Golf   – 50

Angling   – 51

Skiing   – 2

Cricket   – 201

Athletics – 151

Hockey   – **?**

94 What number
should replace the
question mark?

9

7

2

5

1

6

8

1

1

5

8

3

1

2

?

What number should
replace the question
mark in the grid?

2
1
8
9
7
4
2
6
5
4
5
9
3
2
1
7
8
3
?
4

Answer see page 132

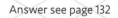

96 A lorry and a van set off from the same point to travel the same journey. The lorry sets off six minutes before the van. If the lorry travels at 60 km/h and the van travels at 80 km/h, how many kilometres from the starting point will they draw level?

97 The sum of each two adjacent squares gives the number above. What number should replace the question mark?

47

?

84   11

60

39

17

Answer see page 132

80

98 Clock A was correct at midnight. From that moment it began to lose three and a half minutes per hour. The clock stopped three hours ago showing clock B. What is the correct time now? The clock runs for less than 24 hours.

Answer see page 132

A   00:00

B   18:50

What should be
the value of the
fourth column?

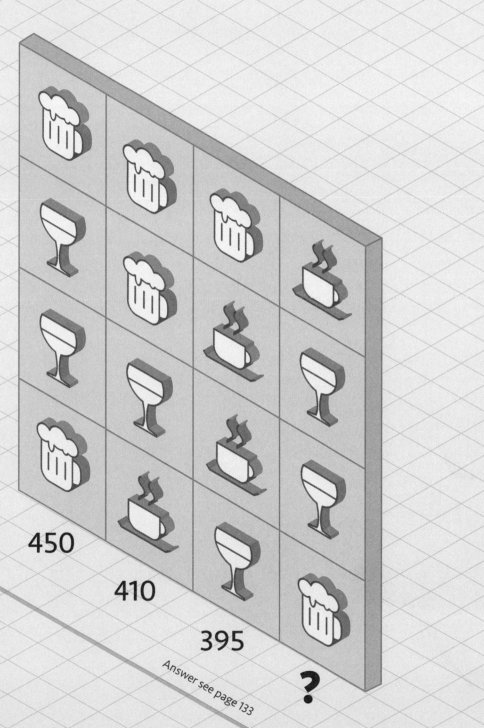

450

410

395

?

Answer see page 133

100

What number should replace the question mark?

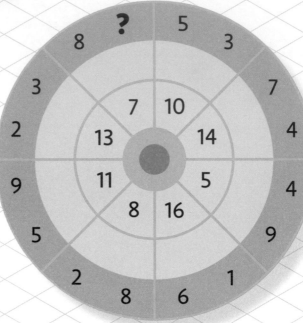

Answer see page 133

Answer see page 133

101

What number should replace the question mark?

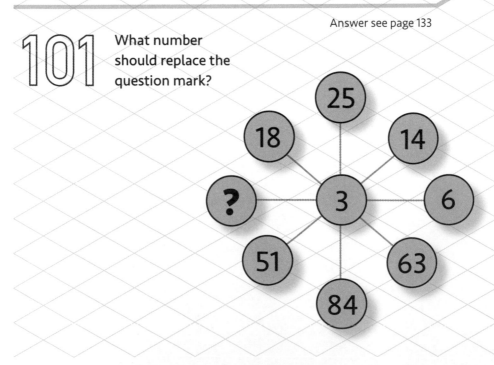

What numbers
should replace the
question marks?

(9) (27) (12)

(16) (64) (20)

(6) (30) (11)

(15) (?) (?)

Answer see page 133

Answer see page 133

The sum of each two adjacent
squares gives the number
above. What number should
replace the question mark?

17

?

9   48

1

22

15

 A 440 yard long train, travelling at 60 mph, enters a one mile long tunnel. How long will elapse between the moment the front of the train enters the tunnel and the moment the end of the train clears the tunnel?

Answer see page 133

Answer see page 134

 What number should replace the question mark in the circle?

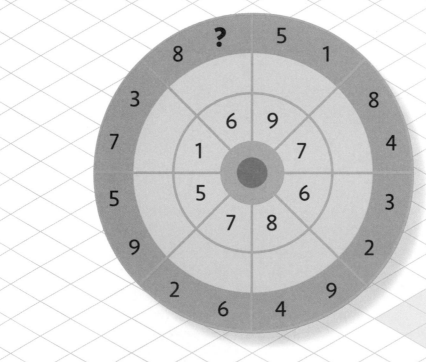

106

What number should replace the question mark?

93   6

32   1

51   4

74   3

85   ?

Answer see page 134

107

What numbers should replace the question marks?

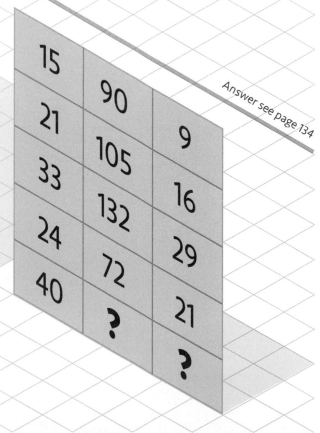

15	90	
21	105	9
33	132	16
24	72	29
40	?	21
		?

Answer see page 134

108 What should be the value of the fourth column?

Answer see page 134

86

61

57

72

?

109

A train covers its outward journey at 110 mph. It returns, over exactly the same distance at 73.33 mph. What is the train's average speed over the entire journey?

Answer see page 134

Answer see page 134

110

What number should replace the question mark?

34   70

15   32

11   24

47   96

29   ?

111 A minibus and a coach set off from the same point, at the same time, to travel the same 140 mile journey. The minibus travels at 50 mph and the coach travels at 35 mph. What will be the difference in their arrival times?

Answer see page 134

112

What number should replace the question mark?

Answer see page 135

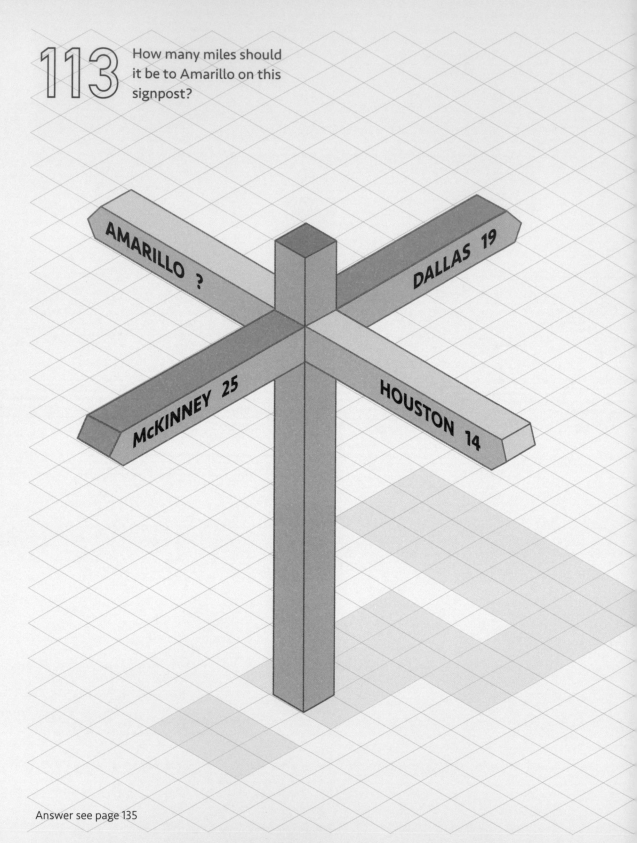

113 How many miles should it be to Amarillo on this signpost?

AMARILLO ?

DALLAS 19

McKINNEY 25

HOUSTON 14

90

Answer see page 135

114

Which is the missing box?

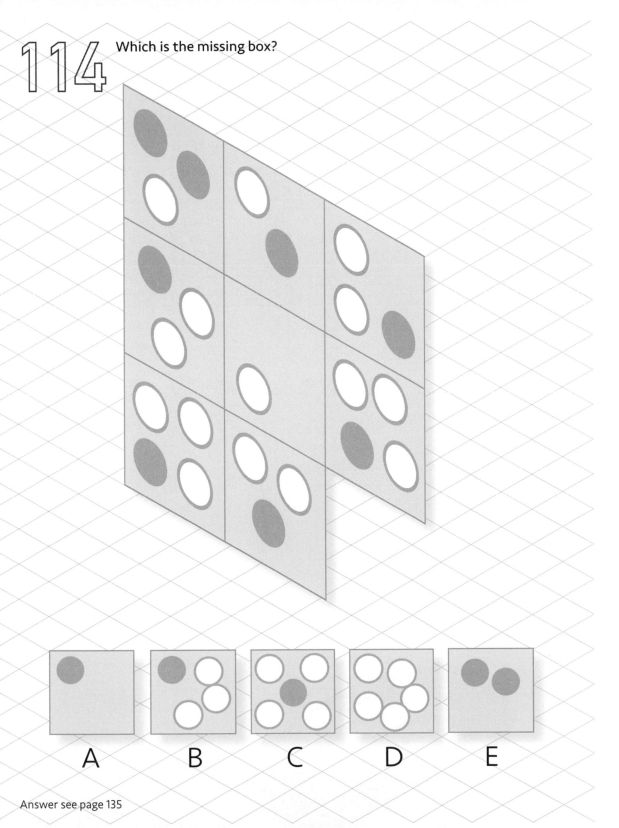

A　　B　　C　　D　　E

Answer see page 135

What number should replace the question mark?

17	8
56	11
32	5
12	3
84	?

Answer see page 135

A car covers its outward journey at 35 mph. It returns, over exactly the same distance at 26.25 mph. What is the car's average speed over the entire journey?

Answer see page 135

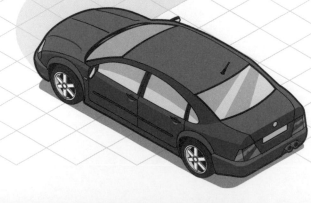

117

Which is the odd one out?

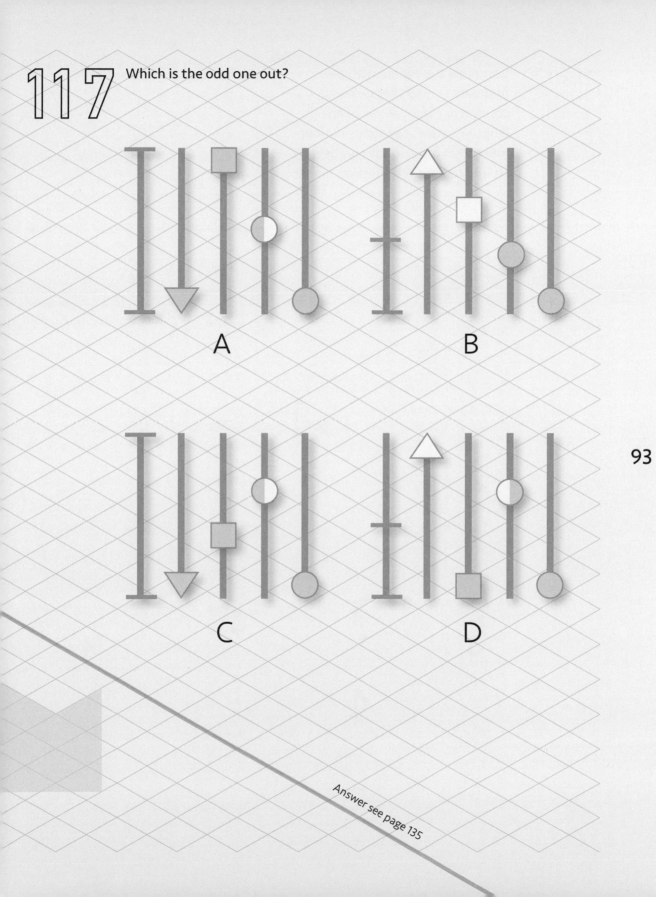

A

B

C

D

93

Answer see page 135

Complete the square with the numbers 1 – 5 so that no row, column or diagonal line of any length contains the same number more than once. What number should replace the question mark?

Answer see page 136

1

2

?

3

5

4

119

What numbers should replace the question marks?

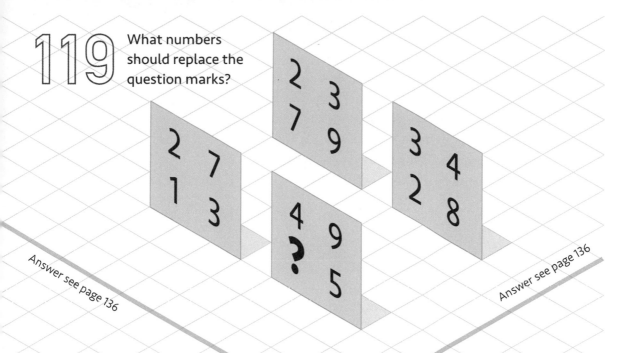

Answer see page 136

Answer see page 136

120

What numbers should replace the question marks?

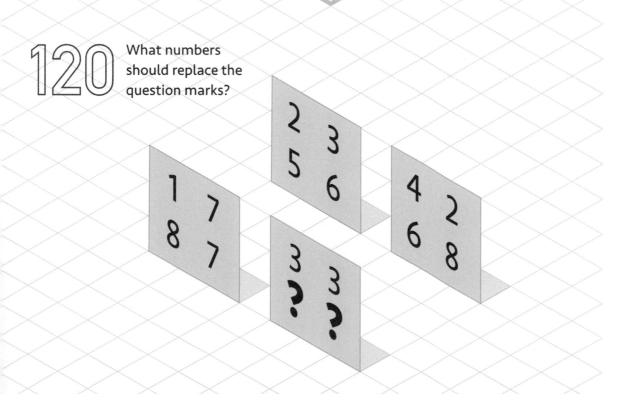

121 At a leisure centre there are three times as many people SWIMMING as there are BOXING. 12 more people are doing AEROBICS than BOXING and 11 less people are playing BADMINTON than doing AEROBICS. 10 people are playing BADMINTON, how many are doing each of the other sports?

Answer see page 136

122 What letter should replace the question mark?

Answer see page 137

ZFN RHB VCP QE?

What number should replace the question mark?

3 5

2 4

1

8 9

1 3

6

5 8

3 4

?

Answer see page 137

124

A customer pays $16.80 in a store. She pays in four different denominations of coins and the largest denomination is $1. She pays with exactly the same number of each coin. The coins that could possibly be available are 1 cent, 5 cents, 10 cents, 25 cents, 50 cents and 1 dollar, how many of each coin did she use and what were their values?

Answer see page 137

125

Car A and car B set off from the same point, at the same time, to travel the same journey. Car A travels at 50 mph and car B travels at 40 mph. If car A stops after 115 miles, how long will it take car B to catch up?

Answer see page 137

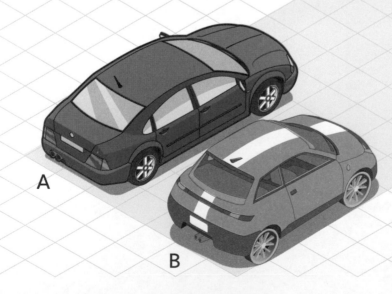

A

B

126 What number should replace the question mark in the grid?

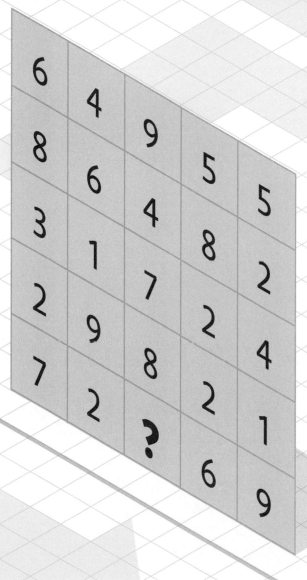

Answer see page 137

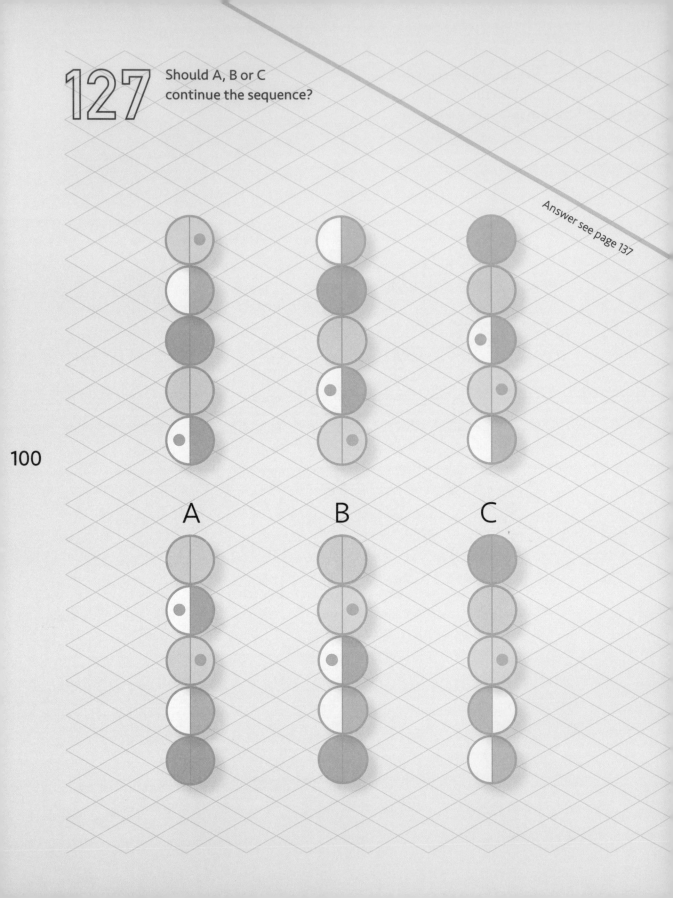

127

Should A, B or C
continue the sequence?

Answer see page 137

100

A

B

C

128 Which number is the odd one out?

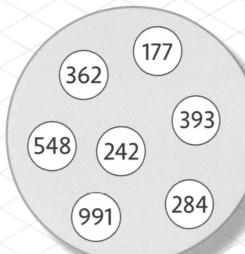

177
362
393
548 242
284
991

Answer see page 138

Answer see page 138

129 Assume you are using a basic calculator and apply the mathematical operations strictly in the order chosen. Replace each question mark with a mathematical sign. Plus, minus, multiply and divide can each be used once only. What are the highest and lowest numbers you can possibly score?

4 **?** 2 **?** 8 **?** 3 **?** 5 = **?**

130 Assume you are using a basic calculator and apply the mathematical operations strictly in the order chosen. Replace each question mark with a mathematical sign. Plus, minus, multiply and divide can each be used once only. In what order should they be used to score minus 42?

$$(2)\ (?)\ (6)\ (?)\ (8)\ (?)\ (3)\ (?)\ (9) = (42)$$

Answer see page 138

131 Which number is the odd one out?

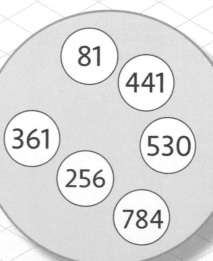

81
441
361
530
256
784

Answer see page 138

132 A 110 yard long train, travelling at 30 mph, enters a three mile long tunnel. How long will elapse between the moment the front of the train enters the tunnel and the moment the end of the train clears the tunnel?

Answer see page 138

Which is the odd one out?

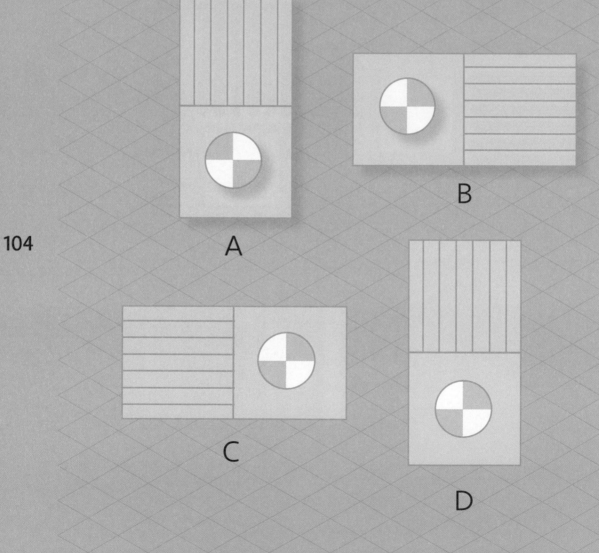

A

B

C

D

Answer see page 138

What number should replace the question mark?

7

3

8 2

2

4 ?

9

5

Answer see page 138

A car has travelled 80 miles at 40 mph. It started its journey with 10 gallons of fuel but its tank has been leaking throughout the journey and is now dry. The car completes 40 miles per gallon. How many gallons of fuel does it leak per hour?

Answer see page 139

What number should replace
the question mark?

A C C A   100

C A C B   107

C B B A   98

B A A C   ?

Answer see page 139

137 Two vehicles set off from the same point to travel the same journey. The first vehicle sets off nine minutes before the second vehicle. If the first vehicle travels at 85 km/h and the second vehicle travels at 100 km/h, how many kilometres from the starting point will the two vehicles draw level?

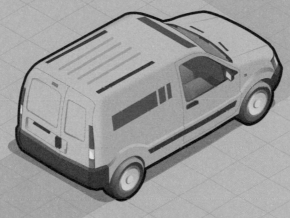

Answer see page 139

A group were discussing where to eat. Four times as many wanted Italian food as wanted Chinese. Five more people chose Indian than chose Chinese and three less people opted for Thai than for Indian. Four people wanted Thai food, how many chose Italian, how many chose Chinese and how many chose Indian?

Answer see page 139

Answer see page 139

108

What number should replace the question mark?

27
19   35

24
18   30

14
10   18

13
11   ?

is to ... as ... is to

A B C

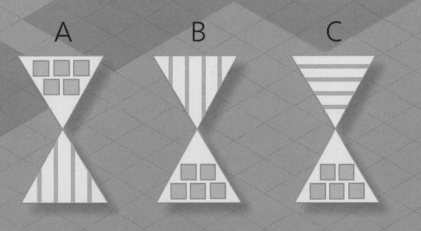

Answer see page 139

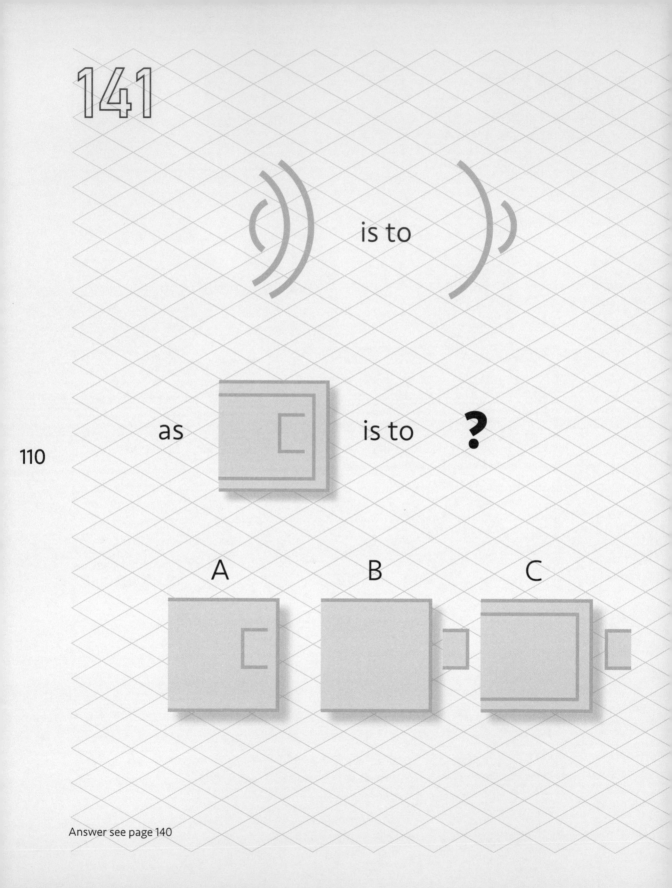

is to

as ... is to ?

A   B   C

Answer see page 140

142

A fire engine travels nine miles to a fire at a speed of 32 mph. Its tank holds 500 gallons of water but has been leaking throughout the journey at a rate of 20 gallons per hour. If the fire engine needs 496 gallons of water to put out the fire, will it succeed?

Answer see page 140

143

Clock A was correct at midnight. From that moment it began to lose one minute per hour. The clock stopped 90 minutes ago showing clock B. What is the correct time now? The clock runs for less than 24 hours.

B

A

Answer see page 140

Answer see page 140

144

In your pocket you have £5.04. It is made up of four different denominations of coins and the largest denomination is 50p. There is exactly the same number of each coin. How many of each coin is there and what are their values?

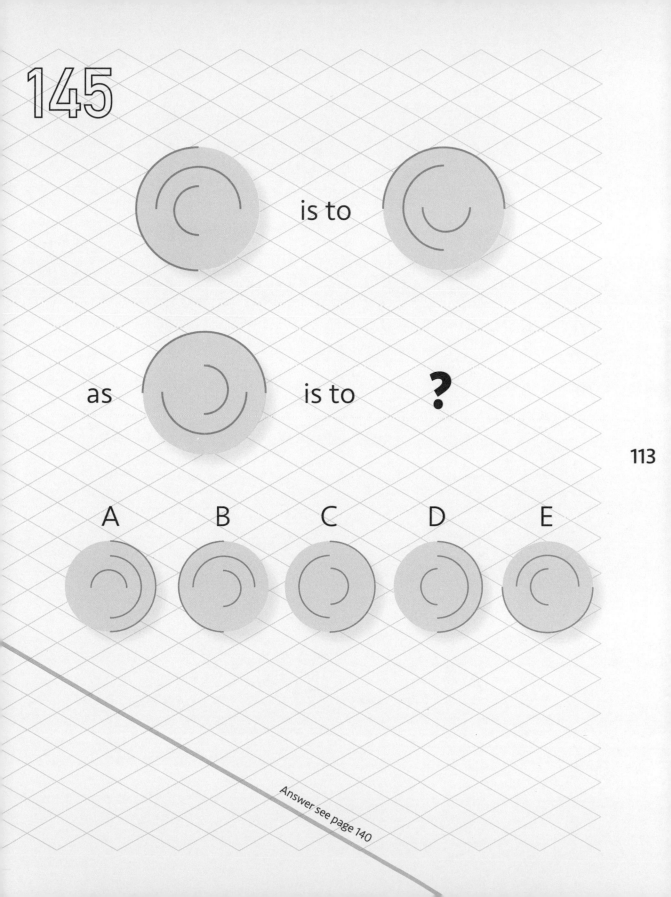

113

Answer see page 140

Should A, B, or C continue the sequence?

Answer see page 140

114

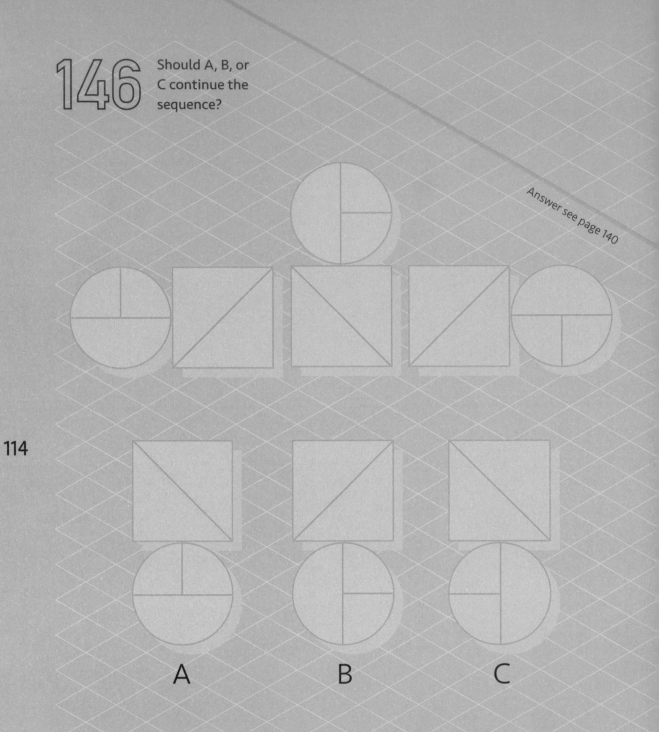

A   B   C

147

A boy is practising throwing a ball through a basketball hoop. In the first 20 minutes he is successful an average of twice every minute. After a further 25 minutes practising, his average stands at seven successes per minute. What is the boy's average number of successful shots per minute of the last 25 minutes only?

Answer see page 141

148

Add together three of the following numbers each time to score 20. Each number can be used as many times as you wish. How many different combinations are there?

Answer see page 141

(2) (4) (6) (8) (10) (12) (14)

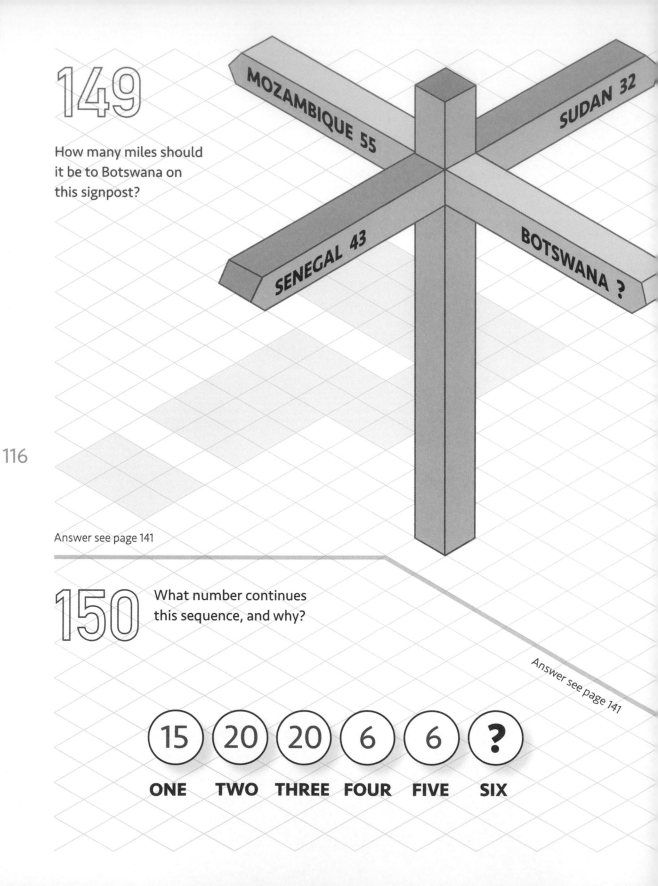

149

How many miles should it be to Botswana on this signpost?

MOZAMBIQUE 55

SUDAN 32

SENEGAL 43

BOTSWANA ?

Answer see page 141

150

What number continues this sequence, and why?

Answer see page 141

15	20	20	6	6	?
ONE	TWO	THREE	FOUR	FIVE	SIX

THE
ANSWERS

04

Multiply, minus, plus and divide.

05

15:30 or 3:30pm.

01

Haiti, Ibiza and Bali.

02

26 minutes and 40 seconds.

06

40 using minus, divide, plus and multiply. Minus 8 using divide, plus, minus and multiply.

03

Five. On each row the first two digit number, minus the next number, gives the last two digit number.

07

Four. Each square totals 21.

08

25. The alphabetical values of the two letters are added together.

09

Eight.

10

23:00 or 11pm.

11

7. Columns total 15, 20, 25, 30 and 35.

12

Vanished.

13

110.

14

4. The gap between the two alphabetical values gives the number.

15

Three. In each square multiply the two bottom numbers to get the two top numbers.

16

6. The number of straight lines in the letters are added or minused as stated.

17

Two. Plus, multiply, minus, divide and minus, plus, multiply, divide.

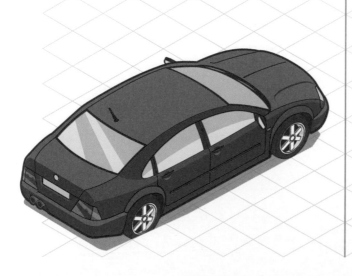

18

444 mph.

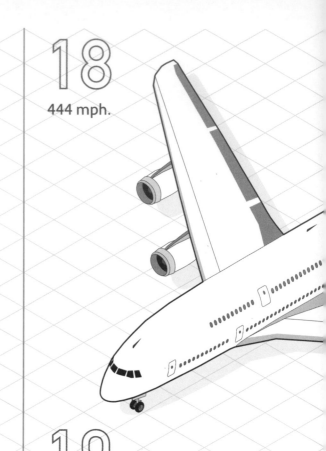

19

1. Add together the alphabetical values of the first two letters and minus the third letter to give the number.

20

7.

21

$4^{2/3}$

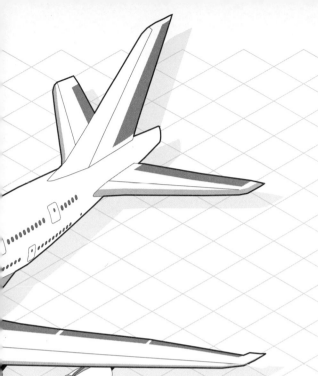

23

No, it will be half a gallon short.

24

120. The numbers are 1x3, 2x4, 5x7, 6x8, 9x11, (10x12), ie., multiplying the next unused odd, or even pair

25

33.

```
                    85
              33         52
         12        21         31
      5         7        14        17
   2        5         9         8        33
```

26

840 mph.

22

Eight.

14 + 4 + 4

10 + 6 + 6

10 + 10 + 2

8 + 8 + 6

14 + 6 + 2

12 + 8 + 2

12 + 6 + 4

10 + 8 + 4

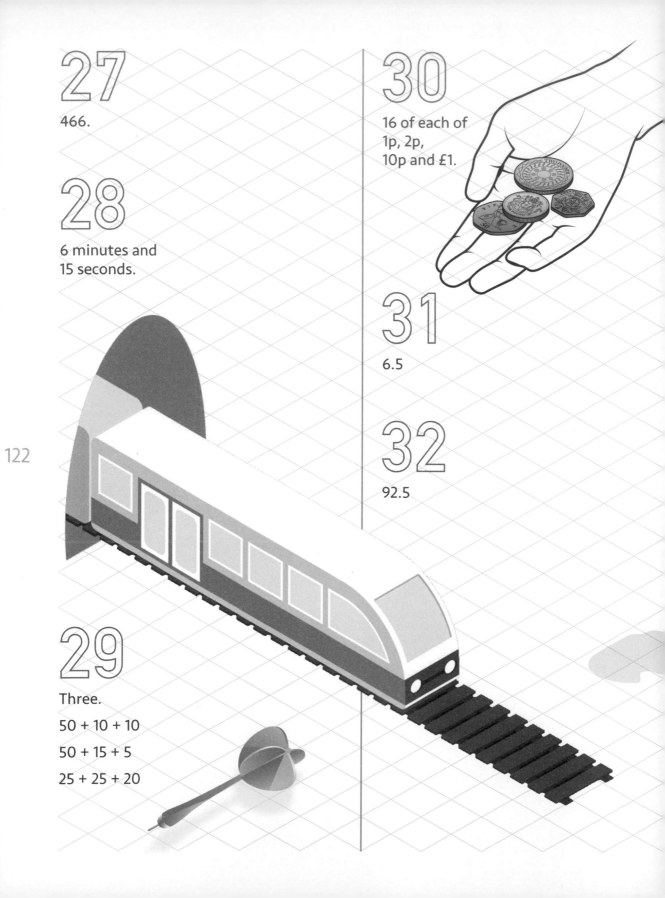

27

466.

28

6 minutes and
15 seconds.

29

Three.
50 + 10 + 10
50 + 15 + 5
25 + 25 + 20

30

16 of each of
1p, 2p,
10p and £1.

31

6.5

32

92.5

33

310. The numbers increase by 20, 40, 60, 80 and 100.

34

Yes with 0.2 gallons to spare.

35

1.25 gallons.

36

73. The two previous numbers are added together each time.

37

E.

38

96.1111

39

24.

40

One. The numbers are divided by 3, then 6, then 9, then 12.

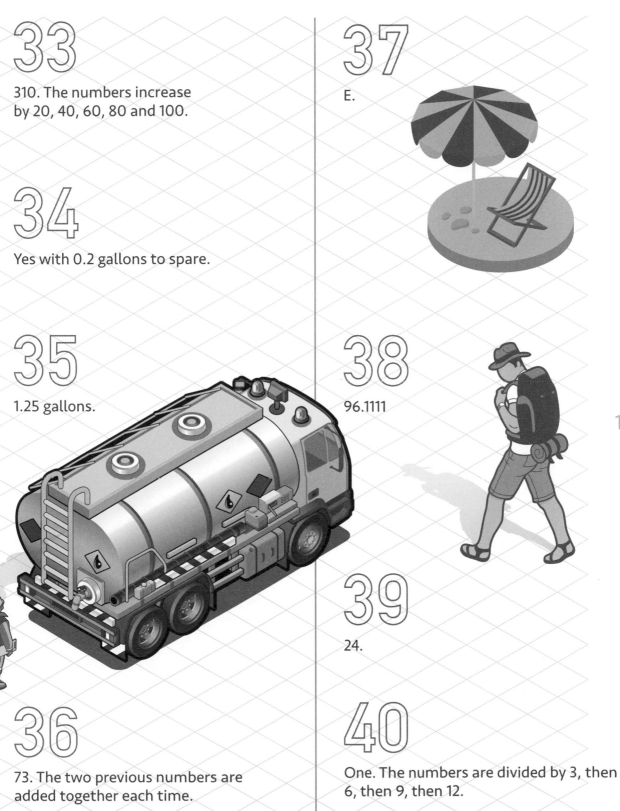

250.

```
                        487
              250         237
         132      118        119
     79       53       65       54
  55       24       29       36       18
```

185.

65 and 18.
Row 1: Box 1 x 20 = box 2,
box 1 + 20 = box 3
Row 2: Box 1 x 15 = box 2,
box 1 + 15 = box 3
Row 3: Box 1 x 10 = box 2,
box 1 + 10 = box 3
Row 4: Box 1 x 5 = box 2,
box 1 + 5 = box 3

44

South. A repeating sequence of arrows pointing south, west, east, north, south, west runs along the top row and returns along the second row and so on.

45

North. A repeating sequence of arrows pointing north, west, east and south runs down the first column and up the next column and so on or, along the top row and returning along the second row and so on.

46

15000. Multiply by 20, then 15, then 10, then 5.

47

2025. The alphabetical value of the first letter gives the first two digits and the alphabetical value of the last letter gives the last two digits.

TURKEY ?

JORDAN 1014

MADAGASCAR 1318

PHILIPPINES 1619

48

4.

49

40. Add 2, add 4, add 6, add 8, add 10 etc.

50

31.

8   13   2

51

28 minutes and 45 seconds.

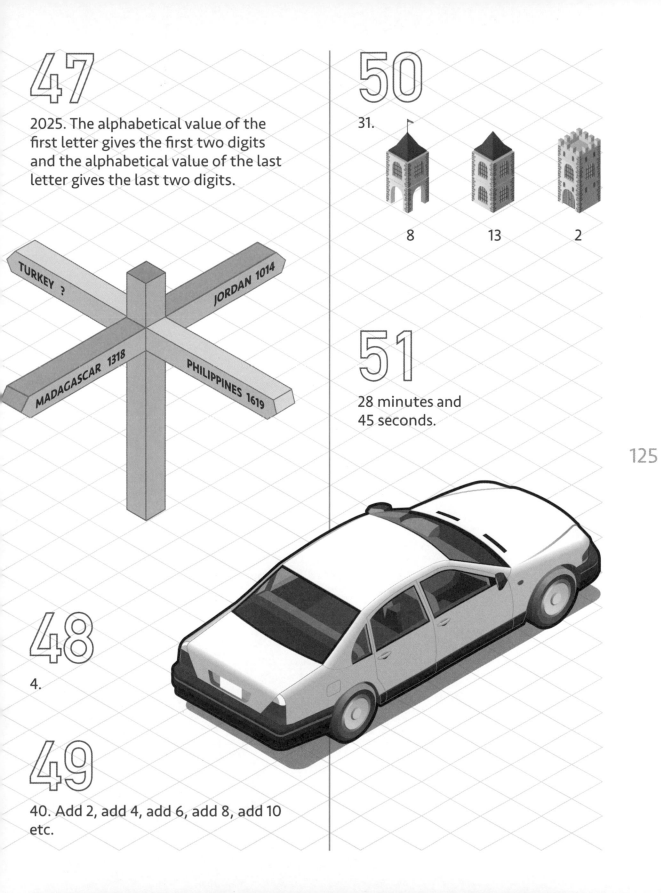

52

One. Moving clockwise along each edge, the alphabetical value of the first letter minus the alphabetical value of the second letter gives the number in the corner.

53

F. The alphabetical values of the two outer letters are added to give the value of the centre letter.

54

Cheese, ham, pickles, pastries, curries, samosas and salads. The reverse order is also valid.

55

88.

```
                    199
              111        88
         62        49        39
    34        28        21        18
24        10        18        3        15
```

56

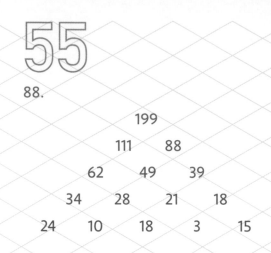

259. Multiply the first and second digits to get the third digit in each group.

57

90. The numbers are 1x2, 3x4, 5x6 etc.

58

125.

59

North. A sequence of arrows pointing north, west, east, south, east and west runs along the top row and repeats along the second row and so on.

60

105.

61

Multiply, plus, minus and divide.

62

Six. Each side of the triangle totals 18.

63

2 miles west.

64

190.

| 10 | 70 | 55 |

65

One third of a gallon.

66

One. In each square add the two bottom numbers to get the two top numbers.

67

19:00, or 7pm.

68

One. Moving clockwise from the top, the sector totals increase by one each time.

69

9 and 24. On each row, divide the first number by three to get the second number and subtract three from the first number to get the third number.

70

$108\frac{1}{3}$ miles.

71

2 minutes and 37.5 seconds.

72

Gym – 12, Swimming – 4 and Walking – 10.

73

190. Each star = 100, each triangle = 50 and each rectangle = 20.

74

Nine of each of 2p, 20p, 50p and £1.

75

496.25.

76

1N on the fourth row.

77

Five.
15 + 5 + 5
10 + 10 + 5
12 + 10 + 3
15 + 6 + 4
15 + 10 + 0

78

18.

10   70   55

79

35. Each number on the left has been multiplied by 3.5 to give the number on the right.

80

11⅓ gallons.

81

Six.

82

Four.
10 + 2 + 2
6 + 4 + 4
6 + 6 + 2
8 + 4 + 2

83

20.

84

Monday
16th
Four
Wednesday

85

Two. On each row the first number plus the second number gives the last number and the third number plus the fourth number also gives the last number.

86

$(17 + 13) \div 6 = \sqrt{25}$

87

#. A sequence of #, %, $, £, ~, X runs down the first column and up the next and so on.

88

200 mph.

89

PAL. The value of each
letter increases by
one with each word.

90

16. The alphabetical values of the
first and last letter are added to
give the number of miles.

91

2 on the top row and 6 on the bottom
row. Add the top row of dashes to get
the third box and multiply the bottom
row of dashes to get the third box.
(1+1 = 2, 3 x 2 = 6).

92

68 miles.

93

100. The total Roman numeral values give the number of viewers.

94

Three. Add the three top numbers to give the bottom two (8 + 3 + 2 = 13).

95

Two. On each row multiply the two outer numbers to give the two centre numbers.

96

24.

97

211.

			474			
		263		211		
	146		117		94	
84		62		55		39
60	24	38		17		22

98

23.00 or 11pm.

99

435.

| 85 | 100 | 125 |

100

8. The sum of the two outer numbers in each sector is placed in the centre of the opposite sector.

101

27. Add the middle number to each number in the top half then multiply by the middle number to get the number opposite.

102

90 and 21. On the top row multiply the first number by 3 to get the second number and add 3 to the first number to get the third number. On the next row multiply by 4 then add 4 to the first number and so on.

103

78.

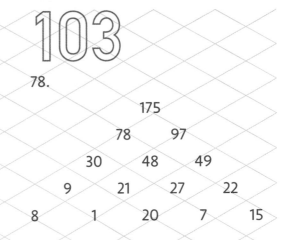

```
              175
          78       97
      30       48       49
    9       21       27       22
  8       1       20       7       15
```

104

1 minute and 15 seconds.

105

7. Opposite sectors total the same.

106

Three. The first digit minus the second digit of the left number, gives the number on the right.

107

80 and 38.
Row 1: Box 1 x 6 = box 2,
box 1 - 6 = box 3
Row 2: Box 1 x 5 = box 2,
box 1 - 5 = box 3
Row 3: Box 1 x 4 = box 2,
box 1 - 4 = box 3
Row 4: Box 1 x 3 = box 2,
box 1 - 3 = box 3
Row 5: Box 1 x 2 = box 2,
box 1 - 2 = box 3

108

46.

20   16   5

109

88 mph.

110

60. Add one to each number on the left and multiply by two to give the number on the right.

111

One hour and 12 minutes.

112

Two. Starting at the top of each group and moving clockwise, add the first three numbers and subtract the fourth number to give the fifth number. (2 + 5 + 4 – 9 = 2).

113

15. The alphabetical value of the last letter gives the number of miles.

114

D. Across each row, (or down each column) dark circles subtract to give box 3, white circles add to give box 3.

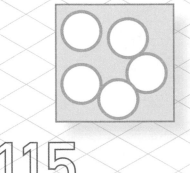

115

12. Add together the two digits of each number on the left to give the two on the right.

116

30 mph.

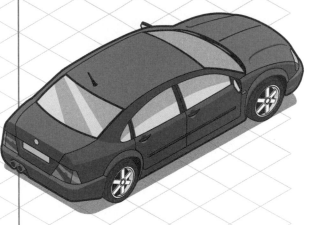

117

B. The triangle and square in the second and third lines of A, C and D have dark shading and the circles in the fourth lines of A, C and D feature dual shading.

118

2.

```
3 2 5 1 4
1 4 3 2 5
2 5 1 4 3
4 3 2 5 1
5 1 4 3 2
```

119

Five. In each square multiply the two numbers on the right to get the two on the left.

120

Six and nine. In each square, the two top numbers are added to give the bottom left and multiplied to give the bottom right.

121

Swimming – 27,
Boxing – 9
and
Aerobics – 21.

122

G. The alphabetical value of the first letter, minus double the alphabetical value of the second letter gives the alphabetical value of the third letter.

123

Four. In each group, the sum of the two numbers on the left gives the top and the sum of the two numbers on the right also gives the top. (5 + 2 = 9, 4 + 4 = 8).

124

12 of each of 5c, 10c, 25c and $1.

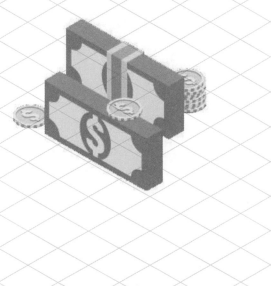

125

34 minutes and 30 seconds.

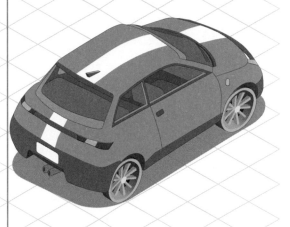

126

Three. On each row, the two digit number on the left, minus the two digit number on the right, gives the centre number.

127

A.

548. For each number, divide the second digit by the first to get the third.

6 minutes and 7.5 seconds.

35 using divide, plus, minus and multiply. Minus 15 using divide, minus, plus and multiply.

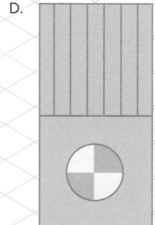

D.

Divide, minus, plus and multiply.

530. All the other numbers are square numbers.

Six, each edge of the triangle totals 20.

135

Four.

136

91. A = 17, B = 24 C = 33.

137

85.

138

Italian – 8, Chinese – 2
and Indian – 7.

139

15. The numbers in the first group rise by eight each time, the numbers in the second group rise by six each time, the numbers in the third group rise by four each time and the numbers in the fourth group rise by two each time.

140

B.

141

B.

142

No it will be 1.625 gallons short.

143

19.30 or 7.30pm.

144

Eight of each of 1p, 2p, 10p and 50p.

145

A

146

C.

147

11.

148

Seven.

12 + 4 + 4

8 + 6 + 6

8 + 8 + 4

14 + 4 + 2

12 + 6 + 2

10 + 8 + 2

10 + 6 + 4

149

53. The number of consonants gives the first digit and the number of vowels gives the second digit.

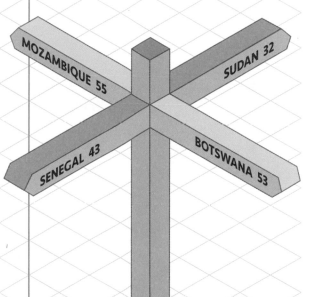

MOZAMBIQUE 55

SUDAN 32

SENEGAL 43

BOTSWANA 53

150

19. The numbers denote the alphanumeric positions of the first letter of the words one, two three, four and five (o = 15, t = 20 etc.) Six begins with s (19).

PUZZLE NOTES

PUZZLE NOTES